Let's Explore Science

PLANTS OUT OF PLACE

COURTNEY FARRELL

ROURKE PUBLISHING

www.rourkepublishing.com

www.rourkepublishing.com

Photo credits: Tad Denson/Shutterstock Images, cover; Ellen Morgan/Shutterstock Images, cover; iStockphoto, cover; Michael Meyer/iStockphoto, cover; Murat Besler/Shutterstock Images, cover, 1 (background); Journal Star, David Zalaznik/AP Images, 1 (top), 19 (bottom); Willey Durden/United States Department of Agriculture, 1 (bottom); Pietro Basilico/Shutterstock Images, 4; Alexander Drack/iStockphoto, 5; Tomasz Gulla/Shutterstock Images, 6; Dorling Kindersley, 7; Shutterstock Images, 8, 9, 32; Diane Picard/Shutterstock Images, 10; Melissa Carroll/iStockphoto, 11; Ricardo De Mattos/iStockphoto, 12; Peggy Greb/United States Department of Agriculture, 13; Troy Maben/ AP Images, 14; Red Line Editorial, Inc., 15, 19 (top), 29 (top), 30 (top), 35, 37, 39 (top); Norman Rees/ United States Department of Agriculture, 16; Tim McCaig/iStockphoto, 17; T. Markley/Shutterstock Images, 18; Danny E. Hooks/Shutterstock Images, 20; Chuck Burton/AP Images, 21; Roger Alford/ AP Images, 22; Alan Marler/AP Images, 23; Mary Ann Chastain/AP Images, 24; Alexander Raths/ iStockphoto, 25; Northwind Picture Archive/Photolibrary, 26; Matt Jones/Shutterstock Images, 27; Willey Durden/United States Department of Agriculture, 28; Ramon Rodriguez/iStockphoto, 29 (bottom); The Brazosport Facts, Miguel Varela/AP Images, 30 (bottom); Pablo Galan Cela/ Photolibrary, 31; Steve Miller/AP Images, 33; Alexandre Meinesz, HO/AP Images, 34; Press Association/AP Images, 36; Keith J Smith/iStockphoto, 38; Christian Février/Nature Picture Library, 39 (bottom); Ken Cote, Indiana Department of Natural Resources, Division of Entomology and Plant Pathology/AP Images, 40; Suzanne Tucker/Shutterstock Images, 41; Duane Ellison/iStockphoto, 42; U.S. Department of Agriculture/AP Images, 43; Andrew Helwich/iStockphoto, 44; Lisa F. Young/ iStockphoto, 45

Editor: Holly Saari

Cover and page design: Kazuko Collins

Content Consultant: Jacques Finlay, Associate Professor, Department of Ecology, Evolution and Behavior, University of Minnesota

Library of Congress Cataloging-in-Publication Data

Farrell, Courtney.
Plants out of place / Courtney Farrell.
    p. cm. -- (Let's explore science)
Includes bibliographical references and index.
ISBN 978-1-61590-322-1 (hard cover)(alk. paper)
ISBN 978-1-61590-561-4 (soft cover)
1. Invasive plants--United States--Juvenile literature. 2. Invasive plants--Juvenile literature. 3. Food chains (Ecology)--Juvenile literature. I. Title.
SB613.5.F37 2011
581.6'2--dc22

                         2010009909

Rourke Publishing
Printed in the United States of America, North Mankato, Minnesota
051111
051011LP-B

www.rourkepublishing.com - rourke@rourkepublishing.com
Post Office Box 643328 Vero Beach, Florida 32964

Table of Contents

WHAT ARE PLANTS OUT OF PLACE?

In the 1800s, salt cedar was brought to the United States from Asia. People thought it would be a nice addition to the U.S. landscape. They also thought the small tree or shrub would help prevent soil **erosion**. But soon, it was causing more harm than good.

Salt Cedar
Field Guide

Appearance:
dense shrub or small tree,
white or pink flowers

Ways it spreads:
wind, water

Removal methods:
hand pulling, herbicides,
biological control

The roots of salt cedar rapidly pull minerals and water out of the soil. Mineral salts collect in the leaves. When the leaves fall, the soil around the salt cedar becomes very saline, or salty. Nearby plants cannot live in the dry salty soil and many die.

Between the 1930s and 1950s, salt cedar spread uncontrollably and affected birds. Before salt cedar spread, it was estimated in one location that 100 acres (40.5 hectares) of land could be home to approximately 150 **species** of wildlife. But after the arrival of the plant, 100 acres (40.5 hectares) could support only four species.

Many plants die when salt cedar invades their habitat.

5

Salt cedar is an example of a plant that is out of place, or a **nonnative plant**. It is also an invasive species, or a nonnative species that grows out of control and harms the **habitat** it overtakes. They are the opposite of **native plants**.

NATIVE PLANTS

In North America, many people consider native plants to be those that existed in North America before European settlers brought plants from their continent.

Today, boats should be checked to make sure any invasive plants are not attached and accidentally transferred to another body of water.

6

The arrows in the food chain point to what an animal eats. In an ocean food chain, plankton may be at the bottom and killer whales may be at the top.

Ocean Food Chain

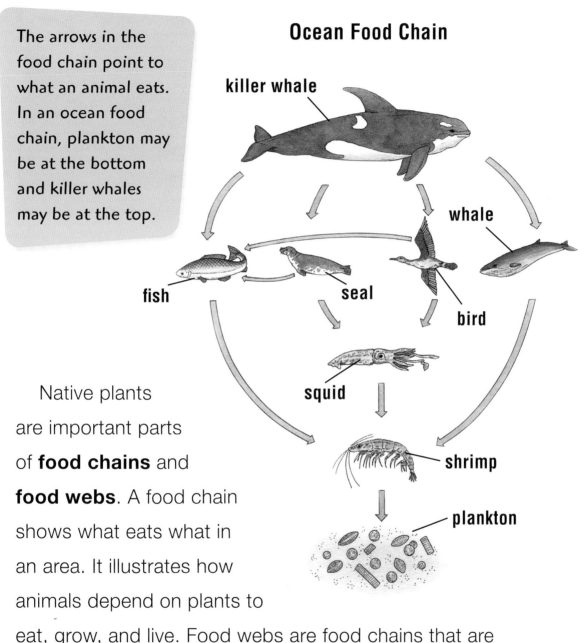

Native plants are important parts of **food chains** and **food webs**. A food chain shows what eats what in an area. It illustrates how animals depend on plants to eat, grow, and live. Food webs are food chains that are connected to each other. Native plants provide food for **herbivores**, which are animals that eat plants. Animals eat the plants' leaves and seeds.

7

DID YOU KNOW?

Plants **reproduce**, or make new plants, in many ways. Some make seeds or spores that grow into new plants after they land in soil. Other plants reproduce vegetatively. This means new plants sprout from the roots or stems of older plants.

NONNATIVE AND INVASIVE PLANTS

When humans move a plant to a new region, it becomes nonnative. It does not belong there. Nonnative plants grow in places other than where they exist naturally.

Nonnative plants are not usually eaten by local herbivores. The animals that ate the nonnative plants in their native region often do not live in the new region. Because many animals in the new habitat do not eat the nonnative plants' leaves and seeds, the plants can spread out of control.

Nonnative plants that spread out of control are called **invasive plants**. Most invasive plants grow and spread quickly. They grow best in disturbed soil. This is soil that humans have affected by digging, building, and other similar activities.

HOW INVASIVE PLANTS ARE HARMFUL

Invasive plants compete with native plants in many ways. Some invasive plants create a lot of shade. Without enough sunlight, certain types of native plants die. A few invasive plants even make **toxins** in their leaves. The toxins poison the soil so other plants cannot live there. After a while, some native plants completely die out.

Some invasive plants pull a lot of water into their roots, making the soil too dry for native plants to survive.

DID YOU KNOW?

Plants grow by converting sunlight into energy. They also need the carbon dioxide in the air. And they need soil, which provides them with water and minerals. Plants put all these things together and make their own food. This process is called **photosynthesis.**

If invasive plants take over, fewer native plants will be around for herbivores to eat. Species such as rabbits and deer may start to die off without this food. These changes to a food chain can also affect top predators such as bears and wolves. Following this pattern, damage from nonnative plants spreads through the **ecosystem**.

When invasive plants negatively affect deer and rabbits, food supplies can decrease for top predators such as bears and wolves.

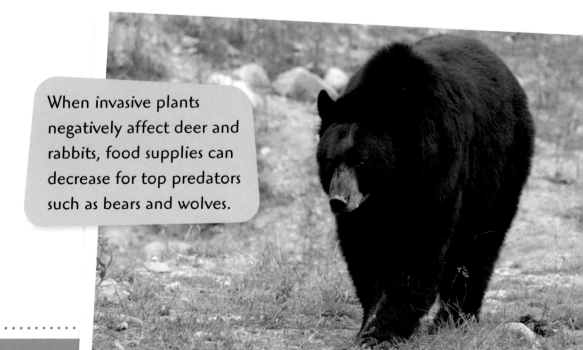

10

Purple Loosestrife Field Guide

Appearance:
green spikes, purple flowers

Ways it spreads:
large number of seeds produced and dispersed

Removal methods:
hand pulling, herbicides, biological control

Not all nonnative plants are invasive. Some grow right where they are planted without spreading too much. For example, many common garden plants are nonnative. In North America, the Siberian iris is a nonnative plant, but it is not an invasive one. It does not overtake other native plants.

Other garden flowers are both nonnative and invasive. They escape into natural areas and spread. An example is purple loosestrife, a tall plant with beautiful spikes of purple flowers. It grows in wetlands all over the United States and **displaces** the native plants.

Native wetland plants are important because they provide food for wildlife. Even though purple loosestrife is pretty to look at, many animals cannot eat it. This nonnative plant damages the food chain.

11

Chapter 2

INVASIVE PLANTS OF THE AMERICAN WEST

The native plants in the Rocky Mountain ecosystem are fragile. Years ago, many of these plants grew well in the region. That changed when ranchers brought cattle to the area in the mid- to late-1800s. The land became overgrazed and bare patches of soil appeared.

CHEATGRASS

Cheatgrass was accidentally introduced to the United States around 1880. Cheatgrass seeds were probably mixed in with a bag of other seeds.

Cheatgrass cheats grazing animals out of food. Like grass, it turns green in spring, but it has already died for the season by summer.

Cattle and horses can only eat cheatgrass for about six weeks before the plant's spiky seed heads form. Animals that eat the seed heads get the spikes stuck inside their mouths. The sharp seeds get under their skin like splinters. The wounds get infected and hurt the animals. The animals' mouths may become so sore that they cannot eat.

Cheatgrass Field Guide

Appearance:
fine, hairy grass that is straw-colored when dry

Ways it spreads:
wind, seeds attached to animals and humans

Removal methods:
herbicides, controlled fires, animal grazing

Cheatgrass causes problems for people too. It catches on fire easily. In areas **infested** by cheatgrass, fires may become more frequent. People who walk in infested fields get the seeds caught in their clothing. Then they carry the seeds and spread them to other places.

One way to get rid of cheatgrass is controlled fires. Controlled fires destroy the plant. Some chemicals can be applied to the plant to kill it. Another way to get rid of cheatgrass is to pull it out by hand or machine. After each plant has been pulled out, a native plant can be put in its place.

Grazing goats are one way to control cheatgrass. The animals eat the plant in the spring before the sharp seed heads form.

Location of Cheatgrass in the United States

present
absent

Cheatgrass is present in all 50 states; however, it is most widespread in the western states.

DID YOU KNOW?

Herbicides can control invasive plants and help clear the way for native species to return. But kids should never handle herbicides! These weed killers are dangerous poisons that can sometimes harm people and wildlife.

15

LEAFY SPURGE

Leafy spurge is a tall plant with yellow flowers and white, sticky sap. It is an invasive weed from Europe and Asia that was accidentally introduced to the United States in the 1800s.

Leafy Spurge Field Guide

Appearance:
bluish-green leaves, small yellow flowers

Ways it spreads:
wind, water, seeds carried by animals

Removal methods:
herbicides, biological control

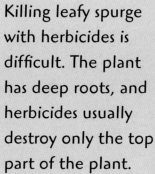

Killing leafy spurge with herbicides is difficult. The plant has deep roots, and herbicides usually destroy only the top part of the plant.

Leafy spurge spreads across rangelands, crowding out plants that livestock can eat. Many grazing animals cannot eat leafy spurge because it makes them sick. Cattle will not eat the grass on fields that are more than 20 percent covered with leafy spurge. Goats and sheep are some of the few animals that can eat it.

Getting rid of leafy spurge is difficult. The plant's roots can go as deep as 21 feet (6.4 meters). The best way to kill leafy spurge is to fence goats or sheep in infested areas.

Chapter 3

THE PLANT THAT ATE THE SOUTH

Some invasive plants do not spread quickly at first. They stay in the area they were planted for decades. If a shift in the climate or a disturbance in the soil happens, they begin to spread uncontrollably.

Kudzu was like this. It lived in the United States for many years without causing problems. Gardeners loved its fragrant purple flowers.

Kudzu Field Guide

Appearance:
green-leaved vine with pink and purple flowers

Ways it spreads:
extremely fast growing rate, seeds carried by animals

Removal methods:
cutting, herbicides, biological control

Location of Kudzu in the United States

present
absent

In addition to invading the South, kudzu has become invasive in two states in the Pacific Northwest: Oregon and Washington.

People planted it on fences and railings. They enjoyed the shade of the leafy vines. Kudzu was even used as food for animals.

Throughout the 1930s, it was planted all over the South to control soil erosion. But by the 1950s, kudzu was climbing telephone poles and trees. Now, it completely covers some old houses.

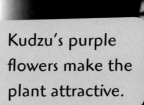

Kudzu's purple flowers make the plant attractive.

19

When kudzu vines climb old forest trees, they block the sunlight the trees need to grow. The trees also bow under the weight of the invading vines. In the southern states, forests are now being smothered. Luckily, kudzu cannot grow in very cold weather, so this is not a problem in the northern states.

Kudzu can blanket forest floors and trees.

Kudzu vines can grow up to ten inches (26 centimeters) per day.

DID YOU KNOW?

Plants can move in several ways from the areas where they were planted. Wind can carry plants' seeds for miles and then drop them in new areas. Rain can wash seeds down sidewalks and streets and then into rivers. Some seeds sprout on riverbanks. Plants can also escape from gardens if birds eat their seeds or fruits. The seeds come out in the bird droppings (along with some fertilizer).

21

Workers try to kill kudzu with herbicide.

22

KUDZU IS HARD TO KILL

Today, people call kudzu *the plant that ate the South*. They joke about closing the windows at night to keep it from growing through the house. But kudzu is a real threat. In 1972, the U.S. Department of Agriculture listed kudzu as a weed and took steps to kill it off.

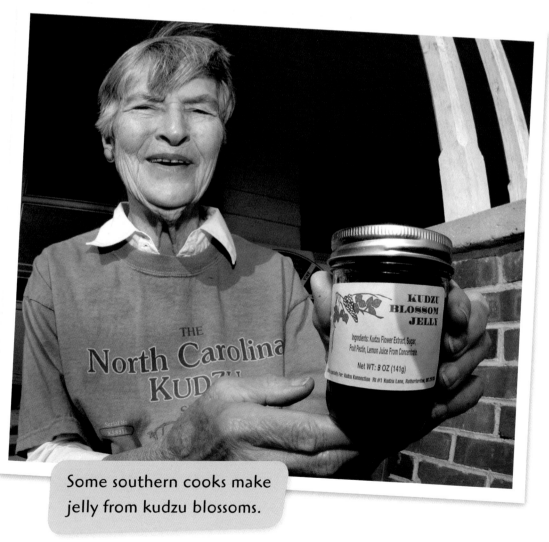

Some southern cooks make jelly from kudzu blossoms.

23

Crafters like to use the rubbery kudzu vines for weaving baskets.

The fight against kudzu has not been easy. The **perennial** plant grows back from the roots every year. If the top of the plant is eaten off by animals or killed by herbicides or cold weather, it will grow back. Kudzu can spread 50 to 100 feet (15.2–30.5 meters) per year. It can take ten years of herbicide spraying to kill one plant. Tons of herbicides have been sprayed on kudzu. But doing this has only slowed its spread. The spraying has not killed it completely.

24

People are exploring other ways to control and use kudzu. Farmers can cut and bale kudzu to feed cattle. Goats and cattle can help control the weed by eating it. This will probably not wipe it out but it may limit its spread. Cooks are experimenting with it as a new food for people in the United States. Researchers are studying kudzu to see if it has medical uses.

Some scientists found that kudzu could be a possible treatment for alcoholism.

Chapter 4

EAST COAST INVADERS

When European immigrants first arrived in the United States, they landed on the East Coast and brought nonnative plants with them. Today, the East Coast is a major shipping center. Nonnative plants slip into the country in ships and planes from overseas. Some of these plants have become invasive. Even nonnative plants that have been brought to the United States on purpose may cause trouble.

When Europeans came to the United States, they brought nonnative plants with them.

The water hyacinth was a colorful addition to aquariums.

TROUBLE IN PARADISE

Nonnative plants can become invasive when people try to do the right thing. For example, aquarium owners sometimes dump their unwanted fish and plants into oceans, lakes, and streams to try to keep them alive. This is a bad idea. The nonnative plants and animals might survive and spread, causing harm to their new habitat.

One aquarium plant that has escaped and grown wild is the water hyacinth. It is now damaging aquatic ecosystems. The water hyacinth clogs waterways, making it difficult for boats to get through. The hyacinths also block the Sun from shining into the water, which ruins the habitat for native plants and fish.

Water hyacinth flowers are often purple.

Categories of Invasive Plants

Category	Description	Example
1 or High	These plants change and harm ecosystems. They crowd out native plants and reduce the number of species that live in an ecosystem.	kudzu, water hyacinth, leafy spurge, purple loosestrife, salt cedar
2 or Moderate	These plants might be troublemakers in the future. They are spreading but have not yet caused severe damage.	tree-of-heaven, meadow knapweed, Japanese knotweed
3 or Limited	These plants are not a problem locally but may be causing trouble in other places.	common forget-me-not, Kentucky bluegrass, hyssop loosestrife

Flooding in Florida has helped the water hyacinth spread throughout the state and into the Everglades.

DID YOU KNOW?

The Everglades region in southern Florida has been damaged by invasive species, pollution, and draining. However, it is still famous for its large number of plants and animals.

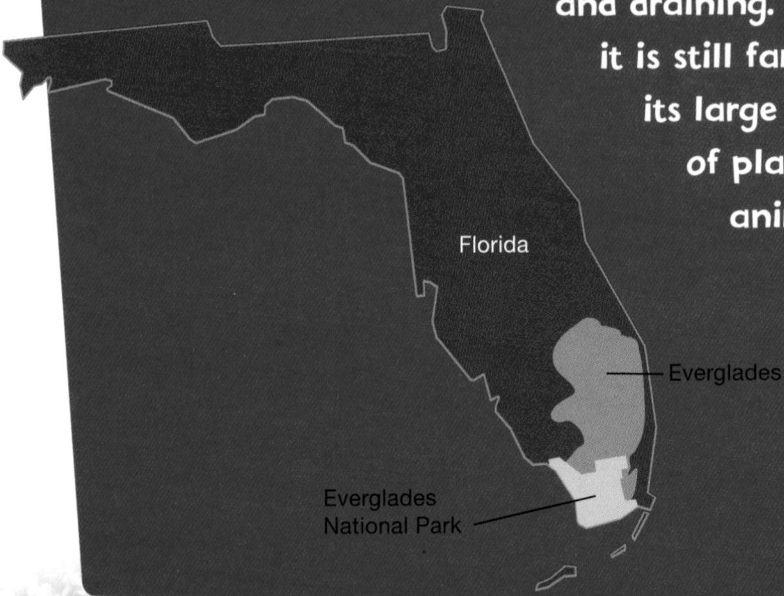

Florida

Everglades

Everglades National Park

Water Hyacinth Field Guide

Appearance:
green aquatic weed with purple flowers

Ways it spreads:
waterways, attached to boats

Removal methods:
pulling by hand and machine, biological control

INVASIVE TREES

Not all invasive plants are vines or weeds. Some are big trees. The tree-of-heaven is originally from China. A gardener first brought it to the United States in the late 1700s. By the mid-1800s, the plant was available to buy throughout the country.

The tree-of-heaven grows quickly and blocks sunlight to native trees. It produces many seeds that rapidly sprout into new saplings, or young trees. It can grow almost anywhere—even in cracks in the sidewalk. In cities the tree's strong roots break up the foundations of buildings. Worst of all, the bark and roots of this tree make a poison that kills other plants.

Tree-of-heaven Field Guide

Appearance:
compound-leaved tree, pale yellow or green flowers

Ways it spreads:
large number of seeds produced, wind

Removal methods:
herbicides, repeated cutting

31

BEAUTIFUL INVADERS

Many invasive plant species are beautiful. People who see them like them at first and let them live. They find it hard to believe that a beautiful plant could be harmful, but some are.

Multiflora Rose Field Guide

Appearance:
oval-leaved shrub with white or pink flowers

Ways it spreads:
seeds carried by animals

Removal methods:
herbicides, repeated burning or cutting

The multiflora rose grows near roadsides and forest edges and in prairies and open woodlands.

The multiflora rose from Japan was first planted as a living fence. Horses and cattle did not like to push past its sharp thorns. Later, the thorny rosebushes started taking over pastures. People and wildlife could not get through. Native plants were displaced.

Chapter 5

INVASIVE PLANTS AROUND THE WORLD

All over the world, plants are finding their way to new habitats. People travel more easily than ever before. When they arrive at new locations, they may be carrying the seeds or spores of foreign plants.

Caulerpa Field Guide

Appearance: marine plant with light-green fronds

Ways it spreads: fragmentation, attached to boats

Removal methods: chemicals including chlorine, covering with black PVC plastic

TROUBLEMAKERS IN EUROPE

One damaged European ecosystem is in Monaco, a tiny country on the Mediterranean Sea. Scientists at the Oceanographic Museum made a mistake that damaged the ocean they love. The museum was using a tropical **alga** named caulerpa to beautify its aquariums. While cleaning the aquariums, someone dumped the water outside. Bits of the alga washed into the ocean and survived. At first no one was worried. Scientists did not think it could survive Monaco's cold winters.

Caulerpa did survive, however. It covered the ocean floor outside the museum in a thick, green mat. It displaced other species of plants.

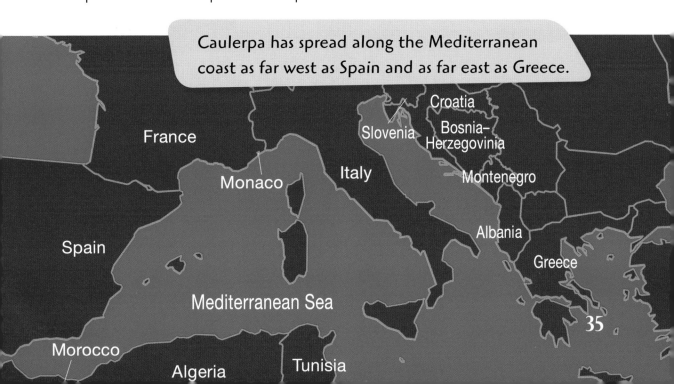

Caulerpa has spread along the Mediterranean coast as far west as Spain and as far east as Greece.

35

DID YOU KNOW?

One of the worst invasive plants in Europe is the Japanese knotweed. It originally came from Japan. This plant can grow nine feet (2.7 meters) tall, and its stems form dense growths called thickets. Pushing through them is almost impossible. Its large leaves create too much shade for other plants to grow.

AN INVADER IN CHINA

Recently the Merremia boisiana has been seen growing up trees in Guangdong Province of China. It has yellow, bell-shaped flowers. Although the new vine is beautiful, it strangles the trees. The Chinese call it the *pretty tree killer.*

The vine is a type of morning glory. It is native to the region north of Guangdong Province. The plant has spread south because humans have moved it and because the weather in the area has become increasingly warmer. The plant does not belong in the south, however. When people cut down the vine, it grows back. Herbicides and burning do not kill it either. The only way to kill the vine is to cut it down and then dig its roots out of the ground.

Location of Merremia Boisiana

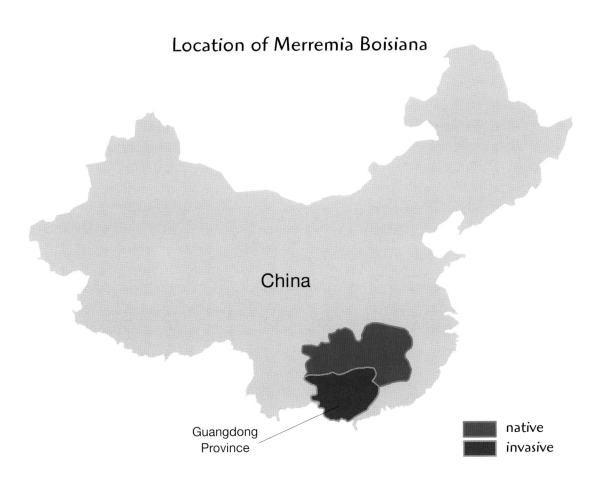

China

Guangdong Province

native
invasive

AN INTRUDER IN THE TROPICS

Tahiti is a beautiful tropical island in the South Pacific. In 1937, the Miconia tree was brought to a botanical garden in Tahiti. The tree's leaves grow up to three feet (0.9 meters) long. Miconia spread through the forests on Tahiti for 44 years before people took action to control it. Now it has taken over nearly 70 percent of the forests, and its large leaves create too much shade for other plants.

Miconia is threatening the forests on other islands as well. It has spread to Hawaii, but the people there are fighting back. They are cutting down the trees and spraying the stumps with herbicides. Teams of volunteers dig up small trees by hand to get rid of them and prevent their spread.

North America

Hawaii

South America

Tahiti

Pacific Ocean

Miconia Field Guide

Appearance:
shrubby tree, leaves green on top and purple underneath, purple or black berries

Ways it spreads:
seeds carried by birds and other animals

Removal methods:
pulling by hand and machine, herbicides

39

Chapter 6

CONTROLLING INVASIVE PLANTS

Invasive plants can harm ecosystems, native plants and animals, and change entire food webs. Most people agree that invasive plants must be controlled. But not everyone agrees on the best way to do that.

Herbicides are widely used to control invasive weeds and trees. But the chemicals that kill plants are also toxic to people, animals, and other plants. People who have been **exposed** to weed killers have a higher chance of getting sick.

40

PESTICIDE APPLICATION
PLEASE KEEP OFF

DO NOT REMOVE UNTIL SUNSET ON _____. FOR ADDITIONAL INFORMATION ON THIS APPLICATION OR ANY FUTURE APPLICATIONS CALL:

Animals and insects are affected by herbicides as well. Rain washes the chemicals into lakes and streams. Fish and insects that live in the water are affected. The animals that eat these fish and insects are affected too.

When patches of invasive plants are sprayed with herbicides, they often die. But so do any native plants that get sprayed by accident. Bare ground is left behind. Invasive plants move back in because they spread best in disturbed soil where nothing else lives. So people spray the area with herbicides again. The cycle continues.

DID YOU KNOW?

Scientists have learned that invasive species are much easier to stop early before they get established. Once a large population of an invasive plant has been established, it may be impossible to get rid of.

ALTERNATIVES TO CHEMICALS

Manual removal is one way to get rid of invasive plants without using herbicides. People remove the plants by hand or with the help of a machine such as a plow. This can be a lot of work. Also, after manual removal many invasive plants come right back. Sometimes manual removal is unsuccessful.

It can take many years of removing a plant before it no longer grows back.

42

Sometimes native plants can be planted where invasive plants have been removed. Doing this helps restore the natural vegetation of the area.

Farmers can have goats and sheep graze in invaded areas. The animals eat the plant and help stop its spread.

BIOLOGICAL CONTROL

Biological control can help control invasive plants. Scientists study an invasive plant in its natural habitat and identify its natural enemies. Those might include insects, fungi, and animals.

Beetles are used to help control salt cedar in New Mexico.

43

Sometimes an invasive plant's biggest natural enemy is an insect. Then scientists put many of those insects into the area where the plant is invasive. Often an insect is brought over from another country. The insect eats the plant, controlling its spread.

But sometimes introducing a new species of insect may start a new problem. The nonnative insect may attack native plants too. Then the insect becomes invasive.

Biological control is being used in many parts of the United States to help prevent and stop invasive species.

Scientists test insects in laboratories. If the insects eat native plants, they will not be good to use for biological control. But if they eat only nonnative plants, they can be put in the wild.

You can help your parents wash their car or boat before leaving a weed-infested area.

HOW YOU CAN HELP

You can help prevent the spread of invasive plants in several ways. For instance, if seeds get stuck on your socks, clothes, shoes, or hair, pull them off in the area where they grow. Taking this step is especially important if you are somewhere far from your home. Make sure you do not bring seeds with you when you return.

Look for bare patches in places where digging and building have occurred. Filling these bare patches with native plants will prevent invasive plants from taking root.

45

Glossary

alga (AL-guh): an organism that lives mostly in water; plural form is algae

biological control (bye-oh-LOG-i-kuhl kuhn-TROHL): controlling invasive plants by using their natural enemies, such as insects

displaces (diss-PLAYSS-es): takes the place of something

ecosystem (EE-koh-siss-tuhm): a community of plants and animals that interact with each other

erosion (i-ROH-zhuhn): the process of wearing or washing away of a substance, such as soil

exposed (ek-SPOZD): not protected

food chains (FOOD CHAYNS): orders of plants and animals in which each feeds on the one below it

food webs (FOOD WEBS): groups of food chains that relate to each other

habitat (HAB-uh-tat): the place in nature where an animal or plant lives

herbicides (URB-uh-sides): chemicals used to kill plants

herbivores (HUR-buh-vors): animals that eat only plants

46

infested (in-FESS-tid): taken over by something

invasive plants (in-VAY-siv PLANTS): nonnative plants that spread out of control, often displacing native plants

manual (MAN-yoo-uhl): done by hand

native plants (NAY-tiv PLANTS): plants that live in the environment where they naturally grow

nonnative plant (NON-nay-tiv PLANT): a plant that moved to or was brought to a new environment where it does not naturally grow

perennial (puh-REN-ee-uhl): a plant that grows again each year without being replanted

photosynthesis (foh-toh-SIN-thuh-siss): the process by which green plants make food by using light for energy

reproduce (ree-pruh-DOOSS): to make offspring

species (SPEE-sheez): one of the groups that plants and animals are divided into

toxins (TOK-sins): poisons given off by plants or animals

Index

Websites to Visit

kids.nationalgeographic.com/Stories/SpaceScience/Invasive-plants

www.sciencenewsforkids.org/articles/20040512/Feature1.asp

www.sgnis.org/kids/

About the Author

Courtney Farrell is a full-time writer who has written nine books for young people. She has a master's degree in zoology and is interested in conservation and sustainability issues. Farrell is certified as a designer and teacher of permaculture, a type of organic agriculture. She lives with her husband and sons on a ranch in the mountains of Colorado.